THE STUDENT'S MUSIC LIBRARY
*edited by Percy M. Young, M.A., Mus.D.*

# THE TECHNIQUE OF
# ACCOMPANIMENT

# THE TECHNIQUE

# OF

# ACCOMPANIMENT

*by*

## PHILIP CRANMER

M.A., Mus.B., Hon.R.A.M.

*Professor of Music in the*
*Queen's University, Belfast*

*London*

DENNIS DOBSON

Copyright © 1970 by Philip Cranmer

First published in Great Britain in 1970 by
Dobson Books Ltd, 80 Kensington Church Street, London W8

Printed by St Stephen's Bristol Press Ltd, Filton, Bristol
SBN 234 77915 2

# CONTENTS

# I

# ATTITUDE OF MIND

'Mr. X WAS the discreet accompanist' or alternatively, 'The discreet accompanist was Mr. X'. So runs the last sentence of almost any notice of almost any song recital.

Discretion is at best a negative virtue; and when our discretion is such that it is noticed by the newspapers, it is probably time we adopted a more positive style of playing. I think that discretion was mainly responsible for the lack of respect given to accompanists twenty or thirty years ago. It was not so much that they were paid badly, or that their names on the programme (if they were there at all) were in the same size of lettering as the name of the printer. It was the general opinion, held alike by other musicians and the general public, that accompanists were an inferior type of musician; insufficiently tooled up with technique to be solo pianists, and certainly not dynamic enough to become conductors, or even chorus-masters.

This state of affairs no longer exists among the best accompanists; they are recognised for what they are—stylish artists and fine players, often considerably better musicians than their partners. At the same time, there lingers a feeling of inferiority among the rank and file; to them this book, and especially this chapter, is addressed.

Most of us accompanists are too retiring and self-effacing, and these characteristics show themselves in modest and obsequious playing. The greater part of our work, however, both with singers and instrumentalists, requires us to be equal partners. There are, it is true, some songs and instrumental pieces in which the accompaniment is to be no more than an unobtrusive background; but in the whole repertory of *Lieder,* of French and English songs of the nineteenth and twentieth centuries, and of opera and oratorio, our rôle is much more positive than this. For a great deal of the time we are not accompanists but solo pianists. In Schubert's introductions we have to establish a mood, often in only a few bars. In Schumann's epilogues we have to summarise a whole song; in *Dichterliebe* and *Frauenliebe und -leben,* indeed, a whole cycle. In Wolf's songs we have to supply exactly the right amount of colour to match the singer's. In Beethoven's and Brahms's violin sonatas we have to match our tone and phrasing with the violinist's. These are but a few examples of our duties. How can we carry out any of them if our attitude is one of inferiority and subservience to the music and our partner? We may quite rightly feel modest and humble beforehand, but if we are to interpret the music artistically these feelings of inadequacy must disappear as soon as we sit down to play. I have never yet seen the direction *con discrezione* on a piece of music.

This subservient attitude can often be observed when the performers come on to the platform. A discreet accompaniment is predictable when one sees—floundering somewhat in the wake of a proud battleship of a soprano, weighed down by three Handel oratorios and four volumes of *Lieder,* so that his gaze can rise no higher than his boots—an insignificant little person who sidles on to the platform and hurries to his stool without even a glance at the audience, much less an acknowledgement of their greeting. He looks hang-dog at his mistress (I use the term in its animal sense), waiting for an imperious nod from her that will send him on his discreet journey through the first Purcell song (with which all the best singers begin their recitals).

If you suspect that you are such a person, then you must do something about it, beginning, as I say, with your entrance on to the platform. You can scarcely do better than follow the advice given to me by the sergeant-major of a training unit I was once at. From a distance of roughly fifty yards he shouted: "You there, don't slouch about like that. Walk along as if you owned the whole bloody world, same as I do." This advice is perhaps excessive for the accompanist, but to heed it will do much to curb our discretion.

To be sure, the assertive accompanist is just as bad. It is not so much that he plays far too loudly (though that is bad enough). Far worse is that he tries to take change throughout, hogging the *rallentandi*, over-emphasising what should be only an echoing phrase, and so on. His approach is very much that of the selfish motorist who would like all others to be cleared off the road. I have heard the opening chord of John Ireland's *Sea Fever* played as if it were the beginning of a piano concerto.

If I have written so much about the wrong attitude, it is because it is much easier to point out the bad than to describe the good. The right approach, which is the most important quality needed by an accompanist, consists in having an open and flexible mind, being prepared to play on successive days two quite different interpretations of the same piece. The accompanist must discuss interpretation on an equal footing with his partner, remembering that his first duty is to the composer and his second to the singer or instrument-alist. He must, in fact, remain artistically honest.

At the same time, he must learn the art of subtle compromise. He must on occasions sink his own musical inclinations. In an ideal performance of a song or a sonata, the audience will have the feeling that the interpretation has been conceived by one composite mind and not two opposing ones. If this aim is always in the accompanist's mind, he will be well prepared to face the technical problems, for the solution of which I give some suggestions in the following chapters.

## 2

# THE TECHNICAL REQUIREMENTS

To play the piano for accompaniment requires no special technique that is not necessary for solo playing; but there are aspects of piano technique which affect the accompanist even more keenly than the solo pianist.

*Sustaining the tone*

The accompanist is almost always partnering a voice or instrument which can sustain a level tone. The voice and the violin and the oboe—indeed most orchestral instruments—can sustain a note for a certain time without any decrease in tone, and can also make a *crescendo* on a single note. The piano can do neither of these things, and so the accompanist (especially in slow music) is always waging a war against the piano's natural inclination towards *diminuendo*. Certainly the solo pianist has this problem too; but I fancy it is more difficult for the accompanist, who is for ever being confronted with a standard of *legato* against which he has to measure the imperfections of himself and his instrument. Any pianist who has played the slow movement of Beethoven's A major violin sonata, op. 30, no. 1, will know what I mean. (There are compensations; for example, no instrument and few singers can match the *diminuendo* that a pianist can make on a last long-held chord.) To create the illusion of sustained tone is the main

11

technical problem the accompanist faces, and this problem will occupy a considerable portion of the present chapter.

Sustained tone on the piano, or rather what passes for sustained tone, comes from a number of factors. First, it comes from smooth, unaccented playing. To play smoothly and to play without accent are almost the same thing. If an organist (whose instrument is incapable of making an accent) wishes to create the feeling of an accent, he shortens the previous note, that is, he plays unsmoothly. More or less the same thing can happen on the piano. If you make a gap between two notes, the second will appear to be accented. And so, if you wish it not to have an accent, you must play the previous note smoothly on to it, that is, with no gap. In the opening bars of Wolf's *Verborgenheit* the F sharps on the fourth and eighth quavers of the bar will always sound accented if there is a gap between the previous Gs and them.

I have two suggestions for smooth playing. The first is mental: try to think of fewer beats, and therefore fewer accents, in a bar. If the time-signature is $\frac{4}{4}$, think of two in a bar; if $\frac{3}{4}$, one. Sometimes the music is too slow for this to be done effectively, but at least make the effort. (This suggestion might well be borne in mind by other musicians than accompanists. Thinking too many beats in a bar is a common reason for stilted and plodding performances.)

My second suggestion is physical: keep the hands

as low as possible over the keyboard; feel that they creep from chord to chord; and when a change of position is necessary, move the hands in a straight line and not in a curve whose descent will produce another accent. I am convinced, after having for some years watched and heard the accompanists of string candidates in Associated Board examinations, that the fault of lifting the hands after each chord is responsible more than any other for unsmooth playing.

Playing smoothly is not the whole solution to this problem. If the tone is to sound sustained, it must be subtly graded from note to note. To state a general rule, long notes must be played more loudly, and short notes more softly. In the above example, if the F sharps are played as loudly as the Gs they will appear to sound louder, because the Gs will to some extent have died away. The F sharps must be played with approximately the same amount of tone as remains in the Gs at the end of their span. In order to preserve the general level of tone in a phrase, it is necessary to make a *crescendo* through the short notes to balance the *diminuendo* that takes place on the long notes, otherwise there is a gradual decline in tone throughout the phrase. The beginning of Schubert's 'Der greise Kopf' (*Winterreise*, no. 14) should sound something like this:

Ex. 2.

though the rises and falls in tone are by no means as steep as they look on paper. The only dynamic marks by Schubert are the *piano* and the two marks up to and away from the top A flat.

A common type of texture requiring smooth playing is one which uses repeated chords in one or both hands. Schubert's 'Who is Sylvia?' and 'To Music' provide examples of this. It is most important to avoid any accents in the right-hand part of these two songs (with the exception that I shall mention in a moment). When you are playing an accompaniment of this sort, try to let your fingers remain touching the keys when they are raised after each chord. It is sometimes, in a very quiet passage, desirable even to prevent the keys from coming right to the top between chords; in this case your fingers will have to remain touching them.

If you wish to have a clear picture in your mind of unaccented repeated chords, listen to the orchestra in the first two bars of 'Comfort ye' (*Messiah,* no. 2). The upper strings play with *louré* (slurred) strokes, in which the bow does not change direction for each note, but stops momentarily, remains on the string, and then continues in the same direction. Another example of this type of chordal texture occurs in the slow movement of Mozart's piano and wind quintet K. 452 at bars 32 *et seq,* where the clarinet, horn and bassoon only just tongue each note. (Ex. 3, p. 15)

Be sure that each note of the chord has an equal weight of tone. Of course, sometimes a note in the chord (usually the top one) requires extra weight. It may also be necessary in this type of texture to play

some chords with an accent. The interludes in 'To Music' and in another song by Schubert, 'Die liebe Farbe' (*Die schöne Müllerin,* no. 16) have a tune at the top of the right-hand chords which needs to be emphasised; and the phrasing in both songs requires that there shall be a slight accent at the beginning of each bar.

### Synchronisation

Making one's playing coincide with the singer's notes (usually referred to as 'following the singer') should not be one of the most difficult aspects of the accompanist's technique, but there are many accompanists who are bothered by it. In any language most words begin with at least one consonant, and this can be used as a warning to play in the next instant, when the singer will have arrived at the vowel and therefore at the note itself. Multiple consonants at the beginning of a word need great care and close attention. Different singers take a different length of time to utter their consonants, and even the same singer will not always utter them at the same speed, but will alter their speed according to the type of song he is singing.

A singer's breathing can occasionally help. A rhythmical singer often gives a good indication of when he is going to begin by his intake of breath. A singing teacher will say that breathing should not normally be audible, but the fact remains that one can sense a singer's intake of breath even if one is not watching him.

Instrumentalists are much more difficult in the matter of synchronisation. They have no consonants to serve as warnings to the accompanist, though occasionally the tonguing of wind-players can act in this way. I have always found the oboe the most difficult instrument in this respect. The amateur player seldom knows exactly when the sound is going to begin. When accompanying an oboist or other wind-player of this sort, begin counting the time from the moment when he makes the attempt to play, and not from when the sound is heard. This may sound facetious, but is meant seriously. I have often been tempted to do otherwise, and have always been reprimanded by the player.

I have known a number of accompanists who were quite certain that their playing synchronised, and it was very difficult to convince them that they were playing before or after the beat. To be behind is a much more common fault than to be in front, though I once heard an accompanist gain a whole verse on a singer during a performance of Herbert Hughes's arrangement of 'The Ballynure Ballad'.* The only cure

* I know this story will not be believed, but I must plead that it is true. The accompanist played five verses (with interludes) while the singer sang four.

for lack of synchronisation is hard practice combined with keen listening.

Look to your synchronisation at all times, but especially (1) when the soloist has passage-work in short notes and you have an accompaniment in longer notes, particularly when these are in block chords, and (2) when the music changes speed.

(1) If the soloist has semiquavers in *allegro* time, he may want to slow up the speed infinitesimally and must be allowed to do so. In Ex. 4 ('Rejoice greatly', *Messiah*, no. 18) play your crotchets with the first of each semiquaver group, listening rather than counting mathematically.

In Ex. 5 (from Handel's violin sonata in E) if you count a mathematically accurate 'two' on the quaver rests you will almost certainly be in front of the soloist when you play the succeeding quaver, because most violinists take minutely longer to play the first semiquaver of each group (or at any rate of the first group in each bar).

In Ex. 6 (from Beethoven's 'Kreutzer' sonata) if you

do not slightly lengthen the left-hand octave Cs at the beginning of each bar, your first right-hand chord in each bar will be early if the violinist leans on his first note in the bar (and many do).

I do not say that I agree with these interpretations, in fact I disagree with the third one; but you must synchronise even when you do not agree.

(2) When the music changes speed, either suddenly (*più mosso* or *meno mosso*) or gradually (*accelerando* or *rallentando*), it is easy to think that one is making the change the soloist wants, while actually making the change one wants oneself. This is especially so at a *rallentando*, where it is tempting to anticipate the soloist's change of speed, or to make a greater change of speed than he does. Remember, too, that *a tempo* takes place as often on the last quaver of a bar as at the beginning of the next bar.

The only time when the change of speed depends entirely on you is when the soloist holds a long note and the accompaniment moves on or holds back. How much change you make will have been decided at rehearsal. You may sometimes have to move on under

a long-held note, not because *accelerando* is marked
but because a singer or wind-player has insufficient
breath to draw out the phrase to its full length. This
needs to be done with great skill and restraint, and
it is sometimes possible to conceal what you have done
even from the singer.

## Pedalling

Good pedalling lies at the root of good accompanying. Yet I have found that by no means all piano
students are certain even of what happens when the
right-hand* pedal is depressed, much less of how they
should use it. Most of them realise that it raises the
dampers from the strings, and therefore causes the
notes that have been played to continue to vibrate
after the keys have come up; this function makes the
sustaining pedal a necessary element in *legato* playing.
Not so many students, however, understand that when
all the dampers are raised other strings which are
harmonically in sympathy with those played are free
to vibrate also. This greatly affects the tone, and most
people would say that it warms, or enriches, it. This
can easily be tested by playing a single note on the
piano, and then playing it again with the same degree
of force, but with the right-hand pedal down. The
note has a fuller sound; and if you can reach across
the piano to touch the string an octave higher than
the one you have played, you may find that you can
actually feel the vibration of the string, though it will
not be as intense as the one you have struck. If you
play a whole chord in these two ways, the difference in
tone is much greater.

* Though it is incorrect to refer to the 'loud' pedal, it is an
understandable convenience to do so.

This richer tone obtained by depressing the sustaining pedal is desirable for the greater part of the accompanist's repertory. It therefore follows that the accompanist's right foot will be down as often as up. Indeed, in really sustained music the foot comes up only to damp the old harmony, in order that two conflicting harmonies shall not sound together. To ensure that the rich tone will not be lost for more than a moment, it is important, having released the pedal for each harmony, to put it down again as soon as possible. This resolves itself into a series of up-down movements of the right foot, and the 'down' should follow the 'up' as quickly as possible. Written out notationally, the first two bars of Schubert's 'Das Wirtshaus' (*Winterreise*, no. 21) would look something like this:

It is not a bad plan to say the words 'up-down' in time with foot-movements.

This technique of putting the hands down and taking the foot up at the same moment comes easily to some. Others find it very difficult, and to them I suggest a simple exercise. Take a hymn-tune and practise it in the following way:

(1) Play the first chord and immediately afterwards put the pedal down.

(2) (The most important stage.) Leave the pedal down and take the hands off the keyboard (look—no hands!). It is vitally important to acquire confidence in the sustaining power of the pedal.

(3) With the pedal still depressed prepare the next chord; and when you are ready, and not before, play the chord and simultaneously release the pedal, letting it come right up. Put the pedal down again immediately.

(4) Take the chord up, leaving the pedal down as in (2), and prepare the next chord.

(5) Continue this practice until you can play the tune smoothly and in time.

The next stage is to practise a number of chordal or simple arpeggio pieces in this way. I suggest Chopin's Prelude no. 20 in C minor, Schumann's Romance in F sharp, Brahms's Intermezzo in E (op. 116 no. 6), Beethoven's 'Pathétique' sonata (slow movement), Schumann's 'A little study' (*Album for the Young,* no. 14).

It will be noticed that all the pieces I have listed were written in the nineteenth century. This is no mere coincidence, because the sustained pedalling I have described will have its greatest use in nineteenth-century music. It will also be useful for most of the twentieth-century repertory. Beware, however, of using it indiscriminately for earlier music. Much music by Mozart and Bach (and eighteenth-century music in general) requires very light pedalling, and often no pedal at all. Do not assume, just because a piano texture is made up of arpeggios, that it must always be pedalled.

Indeed, even in nineteenth-century music there will be occasions when the music is to sound cold or stark, and here the pedal should not be used. Examples occur frequently, and I suggest that there are at least two in Schubert's *Die Winterreise*: one in 'Auf dem Flusse' (no. 7), where in the first verse a completely wrong impression of warmth would be given by pedalling (this should be delayed until the E major section): and the other in 'Der Lindenbaum' (no. 5) at the words *Ich musst' auch heute wandern* (again the pedal must be reserved for the following section in E major). In both these passages Schubert uses the minor key and a stark texture to indicate cold and desolation, and the accompanist must not put his right foot on the heater. I take a third example from Brahms's *Vergebliches Ständchen.* Brahms illustrates, in the accompaniment, the words *So kalt ist die Nacht, so eisig der Wind*; and he uses the minor key and a passage in unison for this. Once again the pedal should not be used.

The right-hand pedal is sometimes used for other than musical reasons, and not least frequently to blur the outline of a difficult passage, in order to conceal bad rhythm or wrong notes. This fault is not exclusive to accompanists, but all (both accompanists and solo pianists) should try to reduce its frequency of occurrence by strenuous and diligent practice.

Turning-over sometimes causes pedalling problems. It is tempting to hold a chord with the sustaining pedal while taking a hand off the keyboard in order to turn the page, but it is sometimes inartistic to do so. Vaughan Williams's *The Vagabond* provides an example. This song has a staccato bass almost throughout, and at the bottom of the first page (bar 12) it is quite wrong to use the pedal to sustain the right-hand chord in order to use the right hand to turn over.

### The left-hand pedal

The action of the left-hand pedal varies from piano to piano. On most grand pianos it moves the hammers to the side so that they do not strike the full number of strings. On most upright pianos it moves the hammers nearer to the strings. On some old pianos it causes the front of the instrument to fall forward and bark the shins of the pianist.

Few general rules can be given for its use. Remember, however, that it is not merely a device for playing more softly, but for modifying the tone. In this it can be compared to the instrumental mute; indeed the same direction, *con sordino*, is used for both, though in modern piano music the term *una corda* is commoner. It is possible, and often necessary, to play softly without using the soft pedal; it is also possible on occasions to play loudly with the soft pedal depressed. On most large pianos it is very difficult to play a real *pianissimo* without using the soft pedal.

The pedal is most effectively used when the piano plays a phrase twice, and the second playing is required to be either an echo or an enhancement of the first. An appropriate moment for its use occurs at the end of

Schumann's *Mondnacht* (bars 61-64, Ex. 8). The marking is my own, and not Schumann's.

**Ex. 8**

# 3

# REHEARSAL

That all those who take part in a rehearsal have already done their individual practice goes, or should go, without saying. If all the time wasted at rehearsal because one person did not know his own part were laid end to end, it would make *The Ring* seem like the 'Minute' Waltz. It always seems as if it is the other players or singers who are guilty, but few of us can claim to be perfect in this matter. I am not referring to sessions at which a singer employs an accompanist to help him to learn his work. Accompanists are delighted to be so employed, and many would be poorer without coaching sessions of this sort. Some accompanists learn as much at them as the singer they are helping.

An accompanist should make every effort to procure the music before a rehearsal. Even if there is no time to practise it, simply to look at music in a train or bus can save time at rehearsal. In the case of songs, make sure of their keys as well as their titles; some songs become almost a different piece of music in another key; do not assume, because you have played a song in one key, that you know it in any key. Occasionally a manageable stretch becomes impossible. Here is a right-hand chord from Schubert's *Im Frühling*.

Try playing it a tone down in F major, in which key many basses sing the song.

At a rehearsal between two or more people it usually happens that one emerges as the leading personality, possibly because he has the most to say (a bad reason), or because he is the best musician (a good reason). The danger is that a dictatorship may be set up, and this is very likely to happen when an experienced and famous singer rehearses with a young accompanist. It may also happen, and this would be far worse, when a young singer rehearses with an experienced accompanist. Ideally, both or all the performers should feel free to take part in the discussion, so that, as I said earlier, the performance will sound as if it has been conceived by a composite mind.

If this is to happen, it is most important for an accompanist to speak his views at rehearsal. Nothing shows up more clearly than an ill-rehearsed or hastily-conceived interpretation. Matters of disagreement or uncertainty must be resolved, and it is vitally necessary to rehearse dangerous corners in the music several times. Do not be content to rehearse them three or four times incorrectly, and only the last time correctly. When rehearsing with a soloist, it is necessary to give close attention to the individual aspects of his performance. Does he speak his consonants slowly or quickly? Does he always speak them at the same speed? How long does he take to breathe, and does he breathe rhythmically before beginning to sing (or play, in the case of wind-players)? Does his bowing-arm give, in its preparatory movement, a clear indication of when he is going to play? Are there parts of his compass where the tone is weak (this may apply to both singers and instrumentalists)? Is he rhythmical?

At rehearsal the experienced accompanist seeks out the answers to these questions almost instinctively. The beginner must more consciously listen for them, if necessary one by one. Synchronisation and balance are the first matters to put right, and of these I have already mentioned sychronisation and shall refer to balance later. Features of style and interpretation come next. In *Lieder,* especially in Schubert, there are many imitative passages in which your phrasing must match the singer's. Listen, too, for contrasts in mood between one verse and another, and do not be content to plough through three or four verses of a strophic song with exactly the same tone and phrasing and pedalling.

In instrumental music it is the ability to match one's playing to the string-player's that is the mark of the good pianist. Here there is an extra difficulty, for not only does phrasing vary from player to player but also bowing within the phrase. To play against the violin's or does not match. I forget which eminent pianist it was who, on being introduced to which eminent violinist, greeted him with the question, 'Do you make the turn?', referring to the trill at the beginning of Beethoven's sonata, op. 96.

With all these and many other questions to be answered, it is not always easy to listen to one's own playing quite so acutely, but this is certainly no less necessary. The listening must never stop, and must be as critical as if you were listening to your worst enemy playing, and were trying to find every possible fault. Above all, do not be resentful if someone criticises you at rehearsal. If you wish to criticise others, phrase your comments as diplomatically as you can. Remarks like 'Are you marked *fortissimo* at that point?' have been known to break up a promising friendship.

During a song-rehearsal spare as many glances at the vocal line as you can to make sure that the singer is singing the correct notes and words. Singers do more performance by heart than almost any other musicians, and it is only natural that mistakes creep into their work. Particularly is this so in such matters as long notes at the end of a phrase, dotted notes, and words. You are being neither friendly nor tactful but stupid and unkind if you do not point out mistakes of this sort to a singer.

Singers cannot always be persuaded to rehearse *encores*. 'If we get as far as that without disaster, we'll be all right', they tend to say. I think they occasionally forget that, whereas they may have chosen an *encore* at least partly because it does not tax them much, to an accompanist it may be a nightmare—and one which may worry him throughout the concert. I agree that singers have the right not to decide which *encore* to sing until they have assessed the audience's reactions; in this case you must ask them to rehearse any likely ones.

The most important problem for you to solve at rehearsal, or to try to solve, is that of balance. The accompanist is never free from this problem, and he

28

never solves it finally. There are no rules to be formulated, because no two singers sing with exactly the same amount of tone, their tone varies from the high to the low parts of their compass, instrumentalists vary, pianos vary, halls and rooms vary, and audiences' absorption of sound varies. In the early stages of your career, you must rely for balance on the comments of friends whose judgement you trust. To be able to rehearse in the concert hall itself is greatly helpful, especially if you can have a third person (not a friend of the singer) to walk round and report from various parts of the hall. Even so, remember that the acoustics of an empty hall may be very different from those of the same hall filled with an audience.

With experience one learns to some extent to gauge a nice balance. It becomes possible to sense from the piano whether a singer's or instrumentalist's tone is reaching the back of the hall or not. I find as a very rough guide that, if from where I sit at the piano the balance is slightly against the soloist, out in the hall it will be right.

However hard one tries to avoid them, there will inevitably be occasions when no rehearsal is possible. In these circumstances the performance can never be entirely satisfactory, and one can only hope that nothing catastrophic will occur. Make sure that you and the singer are performing the same song (I have known the contrary to happen), that your copy is in the right key, and that you know at what speed the singer wants to sing it.

In the matter of rehearsal the competitive-festival accompanist is the worst served of all, and may in a week play up to five hundred items without rehearsal.

There is probably no remedy for this that will not add greatly to the festival's expenses. I feel, though, that the organisers increase the accompanist's difficulties when they push the piano almost off the platform for the vocal classes. If it remained in the middle, the singer and accompanist could far more easily hear each other, thus removing one difficulty.

# 4

# PERFORMANCE

The great thing about a performance, indeed the only thing, is that it is a performance. This simple fact is responsible for all the heartaches, the nerves, the psychological upsets, the tantrums, and all else that makes a performance so much better or worse than a rehearsal or a run-through. For whether a performance is better or worse, the one certainty is that it is different.

I imagine that no two performers' emotions at the beginning of a concert are quite the same. My own uppermost feeling is that everything is quite unfamiliar, that I have never been in the hall before, even if it is a hall I know well. Sitting at the piano feels strange. The music looks unfamiliar (sometimes it is). I sense during the first few chords that I cannot hear properly; the sounds come from afar. Often I have to tense my arm and hand muscles more than I ought, in order to control them at all. In short, I suppose that I am nervous.

I have no cure to offer, nor do I believe that there is one. Neither do I believe that it is right or wrong to be nervous or not to be nervous. It is a matter of personal chance, and each performer must use his own method of controlling his nerves. My only reason for describing a feeling which you have almost certainly experienced yourself, on the field of play, or after a public dinner, or in the headmaster's study, is that an accompanist must bear in mind that his partner may also be suffering in this way. If he is a singer, he will not even have the reassuring sight of the music in front of him.

By your playing of the first few bars of the first song you can do much to restore his confidence. I suggest that you play these bars rather more strongly than you have at rehearsal. If it is a loud song, so much the better; but if it is Purcell's *Music for awhile* (and it probably is), a slightly firmer playing of the bass line can do no harm at all. Moreover, to concentrate on sympathising with your partner's nerves may help you to forget your own.

Once you are under way, assess the acoustics of the hall as quickly as you can, and adjust the balance to suit them. I know that not nerves but unfamiliar acoustics are really the reason for my not being able to hear properly at the beginning of a concert.

Throughout the performance spare a thought for what is coming, in order that passages which you have been rehearsing in detail shall not take you by surprise. It is most demoralising when some small piece of *ensemble* that has taken much rehearsal, and has eventually fitted into place, goes astray at the performance. If something of this sort goes wrong, try to accept it coolly and forget about it immediately. I find that the most annoying thing that can happen during a concert is to make another mistake because I was brooding over a previous one.

Do not hurry between items or movements, especially with a singer or wind-player, who needs time to recover his breath. If you have to begin solo, be sure that your partner is ready before you start. On two quite separate occasions I have heard pianists start the last movement of César Franck's violin sonata before the violinist was ready. On the other hand, we really must do something about oboists who stop for nearly a minute after the first and third movements of an eighteenth-century sonata, with much relief to themselves no doubt, but none to the audience left poised and gasping on the top of a half-close.

I said above that I had no cure for nerves. There are, however, certain precautions which may be taken to prevent them from extending their attacks. Make sure that you have all your music, and that it is in the correct order. If someone is turning over for you brief him fully, showing him any places where it is necessary to turn back, and also any places, at the end of a quiet passage for instance, where he must not turn. It is on the whole better not to turn over for yourself. There are few complete programmes in which you do not have to fake a chord or two, or use too much pedal, in order to turn the pages (see page 23). If the audience once see you in difficulties they become fascinated, and in watching for a recurrence they are distracted from the music. Inside pages can suddenly fly away on a draught of air; mended music can stick together in the warm atmosphere. I once found myself in the next Act during an aria, because half the rest of a much-mended volume stuck to the page I was turning. Even the worst turner is preferable to situations like this; make him sit far enough away to give you plenty of room, and instruct him to turn exactly when you nod. In this connection remember to allow for the period of time between nod and turn.

When you come on to the platform for the last group, either bring the music for the *encores* with you or leave it accessible in the artists' room. Applause sometimes dies very quickly, and you will not be thanked if an *encore* is thwarted because you could not find the music quickly enough.

Your demeanour during the concert depends on your general appoach to accompanying, and about this I have already written. You should neither seem deferential nor domineering. Enter and leave the platform as if it was your firm intention to do so, and not as if you were doing it by mistake and hoped nobody would notice. Try to appear relaxed but not apathetic. If your partner makes an announcement about the items, or about the translation of a song, give it your full attention, even if you have heard it before. In fact, by behaving as naturally as possible in what are somewhat unnatural circumstances you not only steady your nerves, but also begin to have the audience on your side. This is what every artist hopes will happen, for it immensely improves any performance.

# 5

# SIGHT-READING

I know some really good accompanists whose sight-reading is no more than moderate. This would seem to indicate that the ability to sight-read well is not an essential requirement for an accompanist. At the same time, it is such a useful accomplishment that anyone who sets out to be a good accompanist must do a great deal of practice at it. My experience has been that most musicians are more impressed by brilliant sight-reading than by first-rate accompanying. Many even mistake the one for the other. And so, if a young accompanist is looking for engagements, he is far more likely to succeed if he can sight-read well. There is no doubt that it is a great nuisance for an accompanist not to be a fluent sight-reader. It means that a singer can never decide at the last moment to change his programme, or to change an *encore* to suit his audience. Apart from these reasons, it is enormously useful to any musician, accompanist or not, to be able to play through a large amount of music quickly.

The statement that good sight-readers are born and not made is, if not actually untrue, at least an over-simplification. Moreover, it is a statement which is usually made by bad sight-readers as an excuse for not practising. I concede that some can by nature sight-read better than others; but I am absolutely certain that not only the worst but also the best sight-readers can improve with practice. The best sight-readers are likely to get plenty of practice in any event, and so I make some suggestions for practice for those who are less gifted.

It is a great fallacy to believe that it is cheating to play a piece more than once when sight-reading. It is possible to play through a piece twenty times and still be sight-reading it. I don't recommend this as a regular practice, because I think that more good comes from passing on to the next piece. In my view, though, it is stupid not to read a piece more than once, because only by going through it a second or third time is it possible to discover the mistakes one made in the first performance.

Try from the outset to acquire the habit of playing in strict time, however slowly. If necessary go so slowly that, for example, a piece in $\frac{4}{4}$ time has to be counted in semiquavers. The speed is not important but the note-values are. It immediately follows from this that no incorrect note or chord is to be played a second time with a view to correcting it. This is the most important thing of all to remember in sight-reading. More than any other factor it distinguishes the good sight-reader from the bad. The difference between them can be stated briefly thus: the good sight-reader plays in time and is content to let wrong notes go uncorrected; the bad one doesn't and isn't. Therefore, however conscious you are that you have made a mistake, go on. Put the mistake right the next time.

One of the commonest instructions given to learner sight-readers is to look ahead; they are often told that the best accompanists always look as much as two bars ahead. This is nonsense. The best accompanists may do so if the piece is easy and they know it well, but if they are reading a complicated piece they look a very small distance ahead, even though, relying on musical experience, they may *think* ahead. What they have done is to develop the mechanism between eyes and hands to such an extent that the hands' reaction to what the eyes see is almost instantaneous. This, I am sure, is what is generally mistaken for looking ahead; it is most important to develop this faculty.

Try to read more than one note at a time (this applies whether you are sight-reading or not). A large amount of accompaniment texture is in some chordal form or other. Try to read these chords as an entity rather than as three or four individual notes. When the chord changes, the first thing to notice is whether any note remains the same for the next chord. For instance, in practising the introduction to Schubert's *Who is Sylvia?* notice that the first change of chord in the right hand concerns only the bottom note, and the second only the top note, while for the third change the middle note becomes the bottom note and the other two have to be altered. (See Ex. 9). This is

Ex. 9.

(1) (2) (3)

so simple as hardly to need mentioning. The method, however, should be applied to more complicated textures. For example, in Schubert's *Rastlose Liebe* (Peters edition vol. I, page 222), it is necessary first of all to read the right-hand arpeggios as if they were chords, and then to see exactly how each chord is changed from the last one. (See Ex. 10). To see

**Ex. 10.**

arpeggios as chords helps not only in the reading of them but also in the playing of them; it means that the thumb and fingers are over their notes in good time instead of having to make a last-moment dive for them.

Once the principle of reading more than one note at a time is accepted, it can be extended by practice until it is possible to read a very large number of notes at one glance. Ex. 11 shows the first bar of two songs, Schubert's *Liebesbotshaft* and Brahms's *Meine Liebe ist grün*. Each bar contains twenty-six notes, but each bar is made up of only one common chord, major in the first example and minor in the second.

**Ex. 11.**

In spite of all that I have said about looking ahead and reading a whole bar at once and all the rest, the good sight-reader is made only by regular practice. This practice must be consistent and it must be frequent. Ten minutes a day is better than two hours once a week. This is true of all practice, not only sight-reading.

Grade your exercises carefully. Be really modest about this, and start on easy children's pieces. Even hymn-tunes may be too difficult at first. Sight-reading is partly a matter of confidence, and it is easy to undermine your morale by tackling work which is too difficult.

Finally, however good at it you become, never let sight-reading take the place of hard practice before any performance. An intelligent listener can always distinguish between even the most inspired sight-reading and really good playing.

# 6

# TRANSPOSITION

'Play 'Comfort ye' in flats, old boy, would you?'

This remark was made to me, as we went from the vestry into the church for a performance of *Messiah*, by a tenor whom I had never before met, much less rehearsed with. To play 'Comfort ye' in flats is not difficult; but to play 'Every valley' in flats (and you can hardly transpose the one without the other) is not so easy. And what about the Overture?

However, all that is quite another story. I quote the tenor in order to show that to be able to transpose at a moment's notice is as useful as to be able to sight-read. Singers often manage to acquire a sore throat between the final rehearsal and the concert, and they sometimes bring the wrong music or none at all, so that the only available copy is in the wrong key.

In transposing from one key to another, try to think of the music in the new key. For example, in transposing the chord of G in the key of C up a tone, think of G as the dominant of C. C up a tone is D, and the dominant of D is A. At first this may seem unnecessarily complicated, but most pianists who transpose well agree that it is the best method. To be sure, it has little virtue in transposing isolated chords such as the one above; but it is essential when playing a piece of music. It requires some knowledge of harmony, and some ability to play by ear. Do not, however, be alarmed if you have neither of these qualifications, for they can be acquired. In fact, in learning to transpose you begin to acquire them.

The first step is to learn to play common cadences in as many keys as possible, major and minor. Take such progressions as

**Ex.12.**

Dealing with the first example in detail, try to think of it as a harmonic progression in which the bass goes from the dominant to the tonic, the tenor from the supertonic to the mediant, the alto from the dominant to the dominant, and the soprano from the leading note to the tonic. If it is too difficult to think of all four notes in the chord at once, be content with the soprano and bass and add the alto and tenor one by one.

As I said in the chapter on sight-reading, try from the beginning to play in time. This means preparing both chords in your mind before you start playing. Your object should be to play these cadences and others like them quickly in any key in the following rhythm :

When you have mastered this, practise cadences in various styles, such as the following

**Ex. 13.**

Play these in all keys, again making every effort to play in time.

It should now be possible to tackle some hymn-tunes and children's pieces. Hymn-tunes are not always as easy as they sound, because they have two notes in each hand. Simple pieces by Mozart and Schumann may be more appropriate at first. Transpose them a semitone either way, and then gradually move out to larger intervals.

At this point I must confess to having been a little misleading. I have tried to keep your attention upon transposing harmonically. I must now admit that it is not always possible, unless you are a very experienced and good transposer, to do all your transposing in this way. There comes a moment when you have to think of notes and chords at the required distance from the original. This moment will come at different times for different pianists. For some it may come as soon as a piece or song modulates further than the dominant. Others may be able to retain their sense of comparative tonality through the chromatic songs of Wolf and Strauss. When the time comes when you can no longer keep pace with the complicated harmonic progressions, face the fact and be content to read up or down however many semitones it may be. It should, however, always be your aim to transpose harmonically to the full extent of your knowledge of harmony. I write, if not with authority, at least with some experience, for I once had to transpose Wolf's *Der Feuerreiter* for performance. Here the music moves far too quickly for one to think in separate notes and it is essential to think in chords and keys.

When you have decided to transpose by reading the intervals, there are not many short cuts to success other than hard and regular practice. Some cuts are not so short as they might be, either. For example, I have heard the suggestion made that in transposing a piece up a tone it is best to read the treble clef as if it were the alto clef. This seems to me useless advice, because nearly all musicians except viola-players do the converse, that is, they read the alto clef as if it were the treble clef up a tone, making the octave adjustment, of course.

It is quite a different matter when you are transposing a third. Here you can help yourself greatly. If the music is to go up a third, read the treble clef as if it were the bass clef, with a two-octave adjustment; if down a third, read the bass clef as if it were the treble clef. In each case, pray that the other hand will keep in step.

The main argument of this chapter remains, however: think harmonically whenever you can, and your transposing will improve.

# 7

# PIANO REDUCTIONS

Piano reductions of orchestral scores seek to place the orchestral sounds under the span of the pianist's hands. These reductions vary from composer to composer, and even more from editor to editor. The making of them poses the same sort of problem that faces the translator of a play or a poem: to retain the essential sense of the original, and at the same time to make it sound

Ex. 14.

stylish in the new language. Some sort of compromise is necessary in both cases. For instance, how far should a piano reduction of a Mozart operatic aria sound like a Mozart orchestral piece played on the piano, and how far like a Mozart piano piece? The one will retain more of the original, and the other will sound more idiomatic piano music. I show the two extremes at Ex. 14 (bars 7-8 of 'Voi che sapete', *The Marriage of Figaro,* no. 11).

The problem of idiomatic reduction becomes in a sense easier with nineteenth- and twentieth-century music, because the piano as we know it took shape during this period. Bach, Handel, Mozart, and Haydn are more difficult to reduce idiomatically. On the other hand, Beethoven, Wagner, Elgar, and Britten are more difficult from a purely technical point of view. With the latter the problem is how much can safely be left out.

If the maker of a piano reduction can play the piano, he will probably have reduced the orchestral score into a form which is convenient to his shape of hand, and one which his technique can manage. It does not follow that it will be as convenient for others. If a piano reduction is too difficult or has too large stretches for you, you must alter it. The first essential is to play in time, whether for a soloist or a choir. Leave out anything that prevents you from doing this. Octaves, complicated semiquaver passages (especially those in thirds or sixths), large chord-clusters, all must go or be replaced by something simpler if they cannot be played rhythmically and with correct phrasing. The piano reduction of 'Is not His word like a fire?' (Mendelssohn, *Elijah*) begins so in one edition

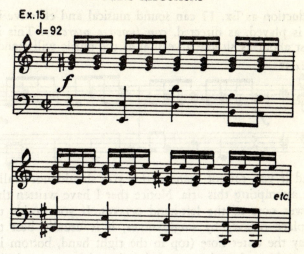

and this pattern, with short breaks, runs through the aria. This is appallingly difficult for the ordinary pianist, and needlessly so. I for one am not going to waste time practising it. If Schubert writes wrong-way-round *tremolos* in his piano accompaniments (and he does so in *Die junge Nonne,* among other songs), we must practise them and try to get them right. But if a piano reduction contains them, we shall at least

Ex. 16.

change them round thus :

If you do not appreciate the distinction between the two, play through 'Halt!' (*Die Schöne Müllerin,* no. 3) and feel the difference at bar 38.

Simplification of piano reductions can go an extraordinarily long way and still retain some of the sense of the original orchestral score. Even as simple a reduction as Ex. 17 can sound musical and effective if it is played, as directed, *con fuoco e marcato*. This is just about as simple a version as the music will stand,

**Ex. 17.**

and if you cannot manage this you should not really be attempting this aria. Notice that I have written the lower note of the left-hand octaves. If you decide to replace octaves by single notes, it is usually best to play the outer note (top in the right hand, bottom in the left), unless the top of the piano has a very shrill, or the bottom a very muddy, tone.

If you are playing a reduction of a work from the *continuo* period (roughly 1600-1750), you will need to exercise great care, because the editor, in addition to reducing the orchestral part, may or may not have realised a *continuo* part in places. The orchestral score should be consulted to find out which in the reduction is orchestra and which is *continuo*. 'Mighty Lord and King all glorious' (Bach, *Christmas Oratorio*, no. 8) is an especially complicated example. The orchestration is trumpet *obbligato,* strings (flute in unison with the first violins), and *continuo*. Two illustrations from it will show what I mean. In the Novello vocal score bars 23-24 are shown on the next page.

I have included the figures from the full score. The chords on the second beats are played by the upper strings, but obviously the *continuo* would play some-

Ex. 18.

thing on the first beat. When you are playing the orchestral part as well as the *continuo,* try to differentiate in your playing between what is orchestral and what is *continuo.* My rendering of these two bars would be something like this:

Ex. 19.

In the same edition bars 30-31 are shown thus

Ex. 20.

In bar 30 the upper strings play the three top parts, but in bar 31 the right-hand thirds are editorial, to be played by the *continuo.* This is entirely acceptable if there is an orchestra, because it is an imitation by the *continuo* of a figure the violins have just played. However, if I were playing this as a piano or organ reduction, I should prefer something like this: where it is quite clear what is orchestral and what is not.

Ex. 21.

When an orchestral passage is rich in colour and texture, it inevitably loses much in reduction for piano. In such passages it is sometimes permissible to use rather more pedal, in order to restore a little of the original richness. This is especially true of nineteenth-century operatic and oratorio scores.

Try to familiarise yourself with the orchestral sounds by listening to performances and gramophone records. It helps greatly when playing a solo passage to know which instrument plays it in the orchestral version. In oboe solos, for example, there will probably be a need for well-defined contrasts between *legato* and *staccato;* with a clarinet the treatment may need to be smoother and more suave. To have in mind the appropriate instrument when you are playing will help you to a more musical and stylish performance, even though your playing may not be vastly different from a technical point of view.

# 8

# SCORE-READING

To be able to read from an open score may not be one of the central parts of an accompanist's equipment, but it can be an exceedingly useful one. In particular, it is essential to be able to play the vocal parts at a choral rehearsal.

*Vocal scores*

In the early stages, and probably later on too, the problem is partly an optical one. It is difficult to read even two staves, if they are a long way apart. Therefore your elementary practice must include playing the soprano and bass parts of a vocal score, omitting the alto and tenor. I suggest some graded exercises taken from Elgar's *The Dream of Gerontius*. The numbers refer to the figures in the score and not to pages.

    A.   Soprano and bass only
        (1)   83 (part II) to 85
        (2)   second bar of 72 (part I) to 73
        (3)   50 (part II) to 51
        (4)   first soprano and second bass from 95 (part II) to 101, and then
        (5)   from 89 (part II) to 95.

Practise also some passages where the bass is printed above the soprano. For example, play the semi-chorus bass and the first soprano (the fifth and sixth staves from the top of the page) from the fourth bar of 75 (part I) to the end of the part.

B.   Hands separately.   Right hand plays soprano and alto, left hand tenor and bass.
   (1)   A, (1) above
   (2)   A, (2)
   (3)   30 (part I) to 32, Play the large stretches as well as you can.
   (4)   third bar of 48 (part II) to second bar of 54. Right hand plays the tenor from 50 for ten bars.

C.   All four parts
   (1)   A, (1)
   (2)   115 (part II) to 116. Learn the semi-chorus and chorus parts separately, and then put them together.
   (3)   125 (part II) to 126. This is in three parts only, but is more difficult. Play the tenor and bass with the left hand throughout, even where the parts cross.
   (4)   B, (4)
   (5)   B, (3)
   (6)   63 (part I) to 64.

Practise all these really slowly, as you would a solo piano piece, until they are perfect in note and time. Then, and only then, increase the speed gradually. Some of the passages are really fast. When you can play C (5) and (6) correctly and expressively you will be in a position to choose exercises from this and other works for yourself.

Your practice of A (4) and (5) should convince you that you must always take in three or four notes at a time and memorise them quickly while your eye roves elsewhere. Do not be too scrupulous about tied and repeated notes. Retain an inner part in one hand as much as possible, even if it means crossing the thumbs.

*Orchestral scores*

The transition from vocal scores to orchestral scores need not be a difficult one. The first essential is to learn the alto clef, which is used by viola-players. Middle C looks like this:

Play through as many viola parts from symphonies and string quartets as you can lay your hands upon. Practise them until you can play an *allegro* movement at the correct speed. This, if it teaches you nothing else, will show what a dull musical life viola-players often lead. You will also discover that they sometimes move up into the treble clef.

Having become proficient with the alto clef, you can begin to read some scores. Begin with simple string-quartet movements, and grade your practice in the way that I have suggested for vocal scores. Play the first violin and cello parts together, do some separate-hand practice (right hand plays first and second violin; left hand, viola and cello), and eventually play all four parts. You will not quite so easily find music that lies well under the two hands. Slow movements and minuets from Haydn provide your best chance, though even there you will often find that you have to play the three lower parts with the left hand. Do not be put off by large stretches; you may freely transpose parts up and down an octave in order to span them.

Next you must learn how the clarinet, horn, and trumpet parts transpose. Detailed information is contained in any book on orchestration*. Briefly, clarinet and trumpet parts in B flat have to be transposed down

a tone. A note that looks like

must be played as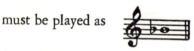

on the piano. Horn in F and *cor anglais* need to be transposed down a fifth, and so what looks like

 must be played as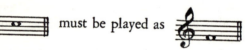

Practise these parts singly in the same way that you practised the alto clef. Practise also the tenor clef

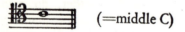 (=middle C)

which is used by bassoons, trombones, and cellos. Later you can combine one of these with a familiar clef, and for this you will find *Preparatory exercises in score-reading* by R. O. Morris and Howard Ferguson useful. When you can read at a glance that

* I suggest Gordon Jacob, *Orchestral Technique.*

(Wagner's *Die Walküre*, miniature score page 894) should be played

you are doing well.

The process takes very much longer to develop than it does to describe. There is a great danger at this point of becoming so involved with transpositions and difficult clefs that you begin to despair of ever learning to read a score properly, and lose sight of its main purpose. The object of orchestral score-reading is to be able to look through a large quantity of music quickly, and even to make some assessments of it. This is invaluable for the conductor, the researcher, the performer, and the teacher. But its principal aim is lost unless the reading has some sense of performance, however slight.

The corollary of this is that at the same time that you are learning the various clefs and transpositions, and practising small details, you must also do quite another type of practice. This consists in playing through a long section of music, possibly a Mozart movement, playing it as nearly as possible at the correct speed, leaving out a great deal, perhaps playing only the first violin and cello parts, perhaps even only the first violin, but trying to grasp what the music will sound like in performance.

Continue these two sorts of practice—the particular and the general—side by side. The particular has to be slow, painstaking, and accurate; the general performed with great dash and many wrong notes, but taken at the correct speed. Over the months they will gradually converge, more and more detail being filled into the one, while the other becomes less slow and takes in more of the general picture.

Choose music for practice as widely as possible. The object of score-reading is to learn a great deal of music quickly. I have mentioned Haydn and Mozart, but there are also symphonies by Boyce, concertos by Corelli and Handel, overtures by Purcell and Lully, all the string-orchestra repertory (Tchaikovsky, Dvorák, Elgar, Vaughan Williams), and a host of other suitable pieces. As a score-reading student you must grab any music in sight, sit down, and play it. You will come to no harm, and, unless you insist on playing the more opulent works of Mahler and Strauss in the early stages, will not depress yourself unduly.

# 9

# CONTINUO

### (with a note on harpsichord accompaniment)

The term *continuo* is short for *basso continuo*, meaning the continuous bass part. It is not an instrument but a part in the full score, played by the bass instruments (usually cellos, basses, and bassoons) and by a keyboard instrument which fills in the middle harmonies according to figures printed under the notes. The term *continuo* is also used loosely to denote the keyboard instrument or its player. This elementary explanation is necessary because many people seem to think that the *continuo* is an instrument. Members of the audience have sometimes come on to the platform at the end of a concert and said to me, 'Do you think I might have a look at your *continuo*?' I have never known quite how to answer this question.

The *basso continuo* had a life of about 150 years from 1600. It is found in all types of music, orchestral, chamber, opera, oratorio, and song. After 1750 it and the harpsichord which usually played the part began to be replaced, in chamber music by the viola or the piano, and in the orchestra by the viola and the horn. This replacement took some years to be completed, and the harpsichord was still in use late in the century.

The main obstacle to good *continuo* playing nowadays is lack of practice. Seventeenth- and eighteenth-century works form only a part of the average singer's or choral society's repertory, and it may be that a provincial accompanist will in a whole season not play more than a score of songs by Purcell and his contemporaries, and perhaps one choral work requiring *continuo*. If he plays arias by Bach and Handel solo (that is, without the orchestra), he is not playing *continuo* but reducing the orchestral accompaniment.

There are three methods of playing a *continuo* part. One is to take a modern edition which has been realised by an editor, and play exactly what he has written. This is a simple method, and in the hands of all but the most experienced players the most musical one, for the quality of music-editing is on the whole high at the moment. Remember, however, that the right-hand part has not been written by the composer, and its importance is therefore limited. Much of the music of this period has an important top and an important bass and not very much of importance in between. If the composer had thought that the right hand ought to have an important part, he would presumably have written one for it; this is precisely what Bach did in his violin and his flute sonatas. The vast majority of instrumental sonatas written before 1750, however, have no such part, and it follows immediately that when playing these works and any like them on the piano, you should be very careful not to use any extra right-hand weight that would be appropriate in a nineteenth-century sonata or song. If any extra weight is required it may well be needed for the bass to compensate for the fact that you will probably have no cello playing with you. On the harpsichord the question does not arise.

This warning is still more necessary if you are using

an edition which is from fifty to a hundred years old. At that time the quality of music-editing tended to be high-handed rather than high, and editors wrote catchy little tunes for the right hand which were quite irrelevant to the music the composer had written. They even altered the bass on occasions.

Ex. 23 shows the beginning of Handel's violin sonata in E major together with the altered bass of one edition (presumably the editor thought he ought to 'correct' the fifths between the two chords).

Another method of playing the *continuo* is to improvise, and about improvising *continuo* parts strong views are held. Most musicians agree that good improvisation is the best method, but what is good improvisation? Here they most certainly do not agree. The question is briefly, to shine or to be restrained? Personally, I have no doubt. I am in almost complete agreement with Dr. Walter Bergmann, whose paper to the Royal Musical Association I recommend you to read (see my suggestions for further reading). *Continuo* playing is a subordinate activity, and is none the worse for being that. Any showing-off or distracting playing harms the music. From the moment when we begin listening for the *continuo* player instead of hearing him as an integral part of the whole sound, we stop listening to the music. I believe that *continuo* playing is the only branch of the accompanist's art where it is not only permissible but also necessary to be discreet. Do not introduce new rhythms and figures which are not indicated by the composer. Imitate the figures of the other instruments by all means, but do not seek to fill in every unoccupied nook and cranny of the bar with interesting wriggles. Do not play notes of a shorter length than the shortest note used by the composer. Occasionally he may write long notes which are rhythmically important; leave them alone. In the solo and chorus 'Thais led the way' from Handel's *Alexander's Feast* the following rhythm occurs frequently

Ex. 24.

I once heard a performance in which the *continuo* player consistently filled in the minim with two more triplet groups. This seems to me entirely tasteless.

If you decide that you do not want to use an edited realisation, and feel unable to improvise, use the third method and write out something beforehand. It is no part of your duty to invent counterpoint on the spur of the moment; this can be left to the experts. Ensure, however, that what you have written is stylish and unobtrusive.

To play *continuo* from a vocal score is not at all easy, because a vocal score is not a *continuo* realisation but a reduction of the complete full score, possibly incorporating a *continuo* part. You must distinguish between the two, and not play anything thematic that is already being played by the orchestra, except of course the bass part. Find out from the full score which is which, and either remember it or write it in your vocal score. On the other hand, there may be places where you should be playing something, but nothing is written in the reduction. (See Exs. 18 and 19.)

In spite of its limited importance *continuo* playing can be a fascinating part of the accompanist's work, and naturally more freedom is justifiable when there are no upper instruments playing. It remains, however, a menial task, and like all menial tasks it is enhanced by being done well. 'Who sweeps a room . . .'

## Harpsichord accompaniment

This is no place for a detailed set of instructions on how to play the harpsichord. There is an exhaustive and excellent article in *Grove* under the title 'Harpsichord playing'. Here I wish only to stress one or two matters which are important to the harpsichord accompanist.

It is *de rigueur* to spread chords on the harpsichord, and the speed of the spread may be varied with the type of movement. Do not ignore the possibility of the downward spread, in which the top note is played first. Whenever and however you spread chords, do it rhythmically. It is not *de rigueur*, when playing in two parts, to play one part in front of the other. Indeed, in most contrapuntal textures any spread should be slight, otherwise the rhythm and clarity are harmed.

It is possible to produce more tone by playing fuller chords. Changes from two notes in each hand to three or four (and back again) may be made within a phrase, and this is the best way to preserve balance with a soloist. Changes of registration are more momentous, and during accompaniment they should normally be used in order to point out structural features in the music, and not to accompany some small *crescendo* or *diminuendo* by the singer or instrumentalist. They are legitimate for echo effects and for important dramatic effects in recitative. However, it is no more musical to make fussy changes of registration than it is to overload a *continuo* part with brilliant *fioriture*. One distracts as much as the other.

# 10

# ORGAN ACCOMPANIMENT

The two special problems of organ accompaniment are those of balance and timing. Both derive from the fact that an organist is usually a great distance from either the singer(s) or the organ-pipes or both. This means that both the sounds he is listening to may come from far away, and not necessarily from the same place. For this reason it is impossible to tell accurately from the console how the balance is between voice and organ. You must ask a third person to listen from the nave. This applies also to timing. The organ sound may take some time to reach the player, for reasons either of distance or of a sluggish action; it may also take some time to reach the singer. In either case, you must play in front of the singer.

## Registration

This depends on so many factors, mainly on the size and placing of the organ, that it is unsafe to give too many general rules. The following suggestions assume that there is no orchestra.

(1) When playing in a long work, or for a vocal recital, try to vary the registration even when the dynamic level is constant. For example (since *Messiah* is the work you are most likely to perform), 'But Thou didst not leave' needs brighter colours than 'Behold

and see'; 'How beautiful are the feet' requires different treatment from 'I know that my Redeemer liveth'. The dynamic level of these four numbers is similar, but it would be dull to play them all on the same stops.

(2) Do not try to copy the orchestration too closely. 'How beautiful are the feet' has a violin solo, but on the organ flute tone sounds better at that altitude. 'The trumpet shall sound' poses a great problem, unless you are lucky enough to have an enclosed Solo tuba. The average Great trumpet stop is much too loud for the bass singer; and most Swell cornopeans are unspeakable when used for solo work. A more practical solution is to use a good diapason as solo, or even to play all the music on one manual at a time. You can then use the Great when the bass is not singing and the Swell when he is.

(3) There will probably not be much opportunity during the choruses for varying the registration, but even on a small organ it is important not to use the full resources for the whole of a chorus, otherwise the climax at the end is lost.

(4) Give the pedals plenty of rest. Even if, in the score, the double-basses are playing, there is no need always to have 16' tone on the organ. To use only the manuals for substantial sections of a movement gives great relief to the ear, and to yourself.

(5) 4' tone helps to make an accompaniment sound interesting, but it may swamp a small voice singing in a low part of its compass. It needs to be used with care, and may have to be confined to interludes, *ritornelli*, etc.

(6) It is unnecessary to change your dynamic level to match every tonal *nuance* of a solo singer. This is

something one does almost without thinking on the piano, but to try to do it on the organ, either by means of stop-changing or by means of the Swell pedal, makes for a fussy and restless performance. Suffer yourself to be slightly too loud or too soft at moments.

## Organ reductions

The accompaniment in vocal scores is usually arranged for the piano, with a few 19th-century exceptions. It will need some alteration before it is fit for the organ. Even so, many of my suggestions in the section on piano reductions may be applied to organs as well. I make one or two supplementary ones.

(1) Unless your pedalling technique is really good, you must often simplify the bass line. Some orchestral bass parts are exceedingly difficult to play on the pedals, not having been written for them. It is justifiable to simplify the part in some such way as is shown in Ex. 25 (bars 10-12 of 'Tis Thee I would be praising,' *Christmas Oratorio*, no. 41). Make sure that the pedal notes are not long enough to blur the semiquaver outline, and do not register them so strongly that the first note of each group sounds like a great thud.

Ex. 25. Manual / Pedal

When the bass part is consistently difficult, as for

example in 'All we like sheep' (*Messiah*, no. 26), it is much better to leave out the pedals completely than to risk playing many wrong notes. Do not play the bass line on a separate manual with 16' tone, because it will not be strong enough to balance the right hand. In this chorus it is entirely musical to begin to pedal at bar 76. In the last of the 'Baal' choruses (*Elijah*, no. 13) the pedals may be omitted until bar 43 of the chorus.

(2) If you play the right-hand part of the vocal score with your right hand and the left-hand part on the pedals, you may be leaving a large gap. Whether this matters or not depends on the texture. If your right hand is playing contrapuntal parts or passage-work, and the chorus has some solid chords in the middle, the gap can remain. See bar 35 of 'For unto us' (*Messiah*, no. 12, Novello, Prout edition), and the corresponding bars later in the chorus. Here you may need both hands to play the semiquaver thirds. If, however, your right hand is reinforcing chorus chords at the octave above, it is better to fill in with the left hand. See bar 14 of 'And the glory of the Lord' (*Messiah*, no. 4). When playing without the chorus, you must either fill in or transpose the right hand down an octave.

(3) When the left-hand part is written in octaves, play the upper note and not the lower note on the pedals. This may be gloomy news to those who pedal with the left foot only, but it sounds far better and helps to reduce the gap I mentioned above. Certainly do not play the lower notes with the feet and the upper notes with the left hand.

(4) It is not wise to play *tremoli* exactly as they are written for the piano. They sound much better on the

organ if at least one note is held down. I should edit
Ex. 15 for the organ as follows

(5) If a pedal passage covers a large range, choose a
suitable moment to change up or down an octave. It is
considerably easier to play the last bars of 'Even so in

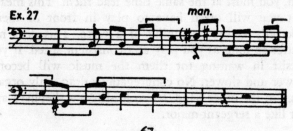

Christ' (*Messiah*, no. 46 or 49, depending on the edition) as shown in Ex. 27 than as written.

I am aware that for seven notes this contradicts what I have said in (3) above, but it is intended only as a makeshift. If you can play what is written, and at the higher octave, so much the better.

*Accompanying a congregation*

I do not wish to trespass for long on matters which are primarily the concern of an organist and choirmaster, but this long-suffering and underpaid type of man is an accompanist for much of his time, and some at any rate of his needs should find a place in this book.

A church congregation needs sympathetic and helpful accompanying as much as any solo singer or choral society. Indeed, it needs a great deal more, because not only are ninety per cent of its voices quite untrained but it almost certainly never has a rehearsal. I am astonished that the one Person to whom most of us are quite content to sing without any rehearsal whatever is Almighty God.

Most of the following suggestions are made assuming that you are not fortunate enough to have a regular rehearsal with the congregation.

(1) Although you are accompanying the congregation, you must at the same time lead them. This means that you will often have to play in front of them, especially at the beginning of a verse. This does not matter so long as you play at a consistent speed. If you persist in waiting for them the music will become slower and slower. No congregations drag, only organists. On the other hand, try to lead like a shepherd and not like a sergeant-major.

(2) Play over a tune at the speed at which you intend to play the verses. There is no other reason for playing over a well-known tune than to show the congregation its speed.

(3) Repeated notes and chords should nearly always be played again; to tie complete chords together merely underlines the unrhythmic nature of the organ. The following sort of performance is most unhelpful to a congregation

Ex. 28.

I wish the organ's capacity for sustaining tone were less than infinite; then organists would be compelled at least to begin each line of a hymn with a fresh chord, and not hold the previous one as in the second bar of Ex. 28.

(4) The type of registration which uses 16' 4' 2', unison off, octave and sub-octave sounds very religious, and is always used by film and television directors for a church scene; but it almost completely hides the tune, and should not be used unless the congregation is really sure of itself.

(5) Two secular *rallentandi* = one ecclesiastical *rallentando;* so runs the organist's arithmetic table. There is no reason why the last verse of a hymn should have considerably more *rallentando* than any other piece of music. In some churches they are still finishing the hymn from last Sunday.

(6) Keep the length of your last chords within bounds. To push in or up all but one stop takes some time if you do them singly; neither does it create a really musical *diminuendo*. When you take your hands off take your foot off too (see the last note of Ex. 46).

# 11

# THE PIANIST IN CHAMBER MUSIC

In this chapter I am going to deal with one or two matters which affect the keyboard-player in chamber music, and I use the term 'keyboard-player' because my first points concern music written before the modern piano evolved.

Bach wrote three sonatas for flute and harpsichord, six for violin and harpsichord, and three for viola da gamba and harpsichord. In all these works he wrote an almost complete right-hand part for the keyboard, and in them the two instruments meet on equal terms. (This is not the case in the works of his contemporaries.) There are a few movements in which one accompanies the other. In the first movement of the fourth violin sonata the harpsichord accompanies (Ex. 29), and in

the third movement of the fifth the violin accompanies (Ex. 30). The third movement of the sixth is for harp-

Ex. 30. Adagio

sichord only. With these few exceptions, the treatment in all these works is that of the trio sonata, with contrapuntal texture almost throughout; and even in the movements for solo with accompaniment, the latter, as can be seen from the two examples, is no mere wallpaper background.

I said that the right-hand part is almost complete. There are passages where it is necessary to do some *continuo* realisation (see Chapter 9). These usually occur when the right-hand *obbligato* is resting. I take an example from the second flute sonata (Ex. 31). The small notes are mine, indicating that the realisation should be as simple as possible to distinguish it from the *obbligato*.

Ex. 31

There do occur also moments where some realisation is necessary even when the right-hand *obbligato* is playing. In the second violin sonata in A major, the last movement is a binary fugal movement, of which I show the beginning at Ex. 32. The Bach-Gesellschaft

Ex. 32.

edition indicates by bracketed figuring that some realisation is necessary. If you do this, it seems logical to play something similar at the corresponding place in the second part of the movement (Ex. 33). In both examples the small notes are mine.

Ex. 33

During the second half of the eighteenth century the harpsichord declined in importance, and it was replaced by the piano. This replacement altered the whole slant of keyboard chamber music. The harpsichord had been one of the least important members of the group; the piano became its most important member. In Mozart's and Haydn's chamber music the piano is often considerably more important than the other instruments. This can be seen most clearly in the piano trios, but it is also evident in some of Mozart's early violin sonatas (which he published as sonatas for piano 'with violin accompaniment'). Even occasionally in the later ones this unequal balance of importance survives. I once heard a comedian raise a great shout of mirth when he announced a concerto for bass drum, and played the following with the pit orchestra:

The opening of K. 481 is hardly less comic if thought of as a violin sonata with piano accompaniment. (Ex. 35)

On the modern piano the problem of balance begins with Mozart, and by and large remains through all piano chamber music of the nineteenth and twentieth centuries. It must be remembered that the first pianofortes had little if any more tone than a harpsichord,

Ex.35

and to deduce from the fact that Mozart knew the early piano that one of his little violin sonatas can be played with impunity on a modern full-size grand piano is too simple. Even in the big sonatas which he wrote later in life, great care and musicianship are required to avoid swamping the violinist. I write with the sounds of a piano-concerto-like interpretation of K.454 by a famous pianist fresh in my ears. The likelihood of drowning the flute or violin in Bach is much less, because of the considerably thinner keyboard texture. This fact illustrates a general point about balance with which I shall now deal.

Balance between the piano and another instrument depends to some extent upon texture. If the piano part is marked *fortissimo* and has extravagant arpeggios, or chords with three notes in each hand, while the string-player is playing at a pitch near one hand or the other of the pianist, there is a balance problem. To solve it the tone of the piano must be moderated, and too much use of the pedal must be avoided, because it is as much the reverberation as the initial percussion of a chord that drowns a stringed instrument. The problem is less severe if either of the following conditions is fulfilled: the string-player's part is at a different pitch from both of the pianist's hands, or the pianist has fewer notes to play.

The last movement of Brahms's violin sonata in D minor, op. 108, has difficulties in balance, though not always in the obvious places. The opening of the movement (Ex. 36) and corresponding points later on

Ex. 36.

are not difficult, because the string-player's part lies between the hands of the pianist, and his scrubbing texture will be audible even if some of the actual notes are lost. You yourself can help by delaying the pedal on each of the long notes until about the second beat.

A more risky spot seems to me to occur at bars

92-93 (Ex. 37), where the violin G sharp is in the middle of the pianist's right hand; the pianist has a large number of notes to play, and may be over-pedalling because of the technical difficulty. In order not to mask the violin, you must make, and continue, the *diminuendo* in bar 92; and to reduce the reverberation use as little pedal as possible (consistent with playing the passage smoothly).

With the cello the problem is sometimes more acute, because the cello does not possess strong tone throughout its compass. Its tone in the middle register is sweet rather than penetrating. I quote Brahms again, in a passage from the last movement of his

cello sonata in E minor, op. 38. This movement begins fugally, and Ex. 38 shows bars 16-17. Here, even though the cello's notes are clear of the piano's, they may easily be swamped by too great a display of energy by the pianist, whose right-hand octaves must be lightened somewhat.

In Brahms the balance of tone often rests upon a knife-edge. His piano-writing can be full-blooded, which may easily lead to too much tone from the pianist. On the other hand, the very fact that it is full-blooded needs to be shown in performance; Brahms will not thank you to play him in an effeminate manner.

A special problem of balance arises in chamber music when the pianist plays the bass part. Playing solo one becomes used to balancing one's right hand with a certain quantity of bass tone, but this amount may not be sufficient if there are other instruments playing. In the following example from Beethoven's trio for clarinet, cello, and piano, op. 11, first movement, bars 101-103, your right hand will probably be

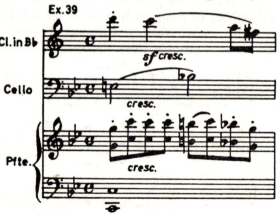

audible without great effort; but you must take care that there is sufficient bass to balance the inner parts of the clarinet and cello.

The next passage (from Dvořák's piano quintet, first movement, bars 61-8) shows an interesting texture in which the cello plays the bass note in some bars and not in others. You should make no more than a slight difference in your left-hand tone between these bars. The cellist can help by ensuring that his tone is not obtrusive, especially in a bar like the seventh one in the example.

Finally, the question of blend. The pianist who plays with one or more string-players is often asked to blend his tone with theirs. It seems to me that 'match' is a better word. I think that to blend implies some sacrificing of individuality, which is something a pianist simply cannot do tonally. He should not even be asked to try. When a phrase is played first by strings and then by the piano, nobody is going to mistake the tone of one for that of the other. Even if

79

this were possible it would certainly not be desirable, because it is the difference in tone-quality which provides the musical interest. What is required is that the playing should match. I have already said a little about this in its application to phrasing (see page 27). Control of the pedal is necessary, too, in order that notes which should not do not run into each other. To

take an obvious example, if the violinist plays an *arpeggio* figure, such as the one in Ex. 41 from Beethoven op. 30, no. 3, last movement, and the pianist has just played the same figure using a great deal of sustaining pedal, the performances will not match.

Control of the length of notes is equally important. I am not referring to dotted crotchets and the rest, important though they are. I mean that all the shades of length between *legato, mezzo staccato, staccato,* and *staccatissimo* must match. This is especially true in fugal passages such as those occurring in the third movement of Brahms's, and the last movement of Dvořák's, piano quintets. Increases and decreases in tone must match from phrase to phrase and within the phrase. In Ex. 42 (Brahms piano quintet, first movement, bars 242-3) the matching of the piano and the strings must include the length of semi-quavers, the amount of increase in tone on to the dotted quavers, and the length of the quavers. The right-hand sixths are awkward to finger and therefore difficult to control, whereas the violins' thirds are comparatively easy.

Ex. 42.

└Vlns. I & II, Vc.────┘  └Piano────────┘

I quote another sort of matching at Ex. 43 (Dvořák, piano quintet, first movement, bar 248).

The string-players stop the bow, and therefore the sound, somewhere about the first of the double-dots. This means that there is a gap in the sound between the long notes and the semiquavers. You must ensure, by listening to your pedalling, that there is a similar gap in the piano part.

String-players can help greatly in this matter of matching, and they can do so especially in two ways. Just as the sustaining pedal is an equipment foreign to them, so *vibrato* is impossible to the pianist. For them to play a phrase with a wealth of expressive *vibrato*, only to hear it echoed 'straight' by the pianist, exaggerates the difference in the nature of the instruments. And so, when playing sonatas and quintets with piano, they should perhaps make their *vibrato* a little less intense than if they were playing a romantic string quartet.

The second way in which they can help is in the matter of the up-bow. I think that a player who persists in making a *crescendo* on every up-bow is an unmusical player with or without a piano, but the fault is aggravated when a piano is playing, and emphasises the difference between the two instruments in the same way that excessive *vibrato* does.

It cannot be doubted that the really sublime pieces of chamber-music are written for strings only; and this fact, together with the thought that you are joining a group who may have played together for a considerable time, is bound to induce a slight feeling of inferiority in your mind when you play piano quintets. I can assure you, however, that the experience of playing these and sonatas with string-players is an exceedingly rewarding one, to be sought as often as possible.

# 12

## PLAYING FOR CHOIRS AND ORCHESTRAS

Playing for a choir is far different from playing for one singer. Among many problems which it sets, the most difficult is that of watching a conductor as well as playing and listening. The first requirement, therefore, is to be able to see him easily. It is vitally necessary to be in such a position that the angle of sight between your music and the conductor is as small as possible. With a grand piano this is easy whether he is higher than you or on the same level (he is hardly likely to be lower). You simply look over or round the music. With an upright piano it is still easy so long as the conductor is sufficiently higher than you. You can still look over the music. The difficulty arises when the conductor is on the same level as you. The best position is one from which you can just see him round the corner of the piano. Insist on having the piano as in the diagram.

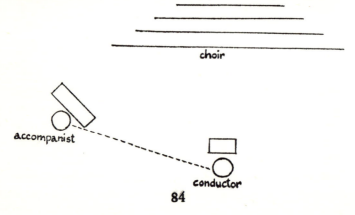

choir

accompanist

conductor

When you are accompanying a conductor and choir, it is important that nothing in either your demeanour or your playing should indicate that you would prefer to be in charge yourself, or that you think yourself musically superior to the conductor. You may think this; it may even be true; but nothing will do the music more harm than for the choir to sense a musical tug-of-war betweeen conductor and accompanist. In order to avoid this, act scrupulously upon any instructions the conductor gives. Never make any suggestions during the rehearsal unless you are asked; keep them for afterwards. It is particularly infuriating for a conductor when, after he has stopped the music in order to rehearse a passage, the accompanist suggests the point of re-start. Always show the keenest interest in what the conductor says, and do not let your attention wander. Try to guess where he will go back to when he has finished talking, and play the choir's notes as soon as possible. Your task will be much easier if (and this will usually be the case) the conductor is a more experienced and better musician than you, because you will be learning at the same time as the choir.

If you are playing music written for choir and piano, play what is written in the accompaniment unless you are asked to play the voice parts. If the music is for choir and orchestra, the suggestions I make in the chapter on reading piano reductions may be appropriate. In particular, it may be difficult for a choir to hear the piano during rehearsal, and it sometimes helps to play the right hand an octave higher and/or the left hand an octave lower (these can often assist the choir when it has problems with pitch). In any case, try to give the choir as clear an indication of the orchestra part as you can. Many choirs can no longer afford an orchestra for their accompaniment, and those who still can will probably not have more than one rehearsal with the orchestra before the performance. The orchestral sounds will in any case be unfamiliar to them, and much can be done by the piano accompanist to help them over this difficulty. The duty of a piano reduction is to reduce; but sometimes editors overdo this and leave out important orchestral leads. Steal a look at the full score if you can, and surprise everyone by putting the leads back.

One of the most important duties of a choir accompanist is to play the voice parts when asked to. This means being able to read an open score (see Chapter 8). There is, however, one important matter, which may sound almost too elementary to mention: the tenors sing an octave lower than their part is written. This may be indicated on their stave by a double treble clef 𝄞𝄞 , or by 𝄞

or it may not be indicated at all. Few things give away a learner accompanist more clearly than playing (for example)

Ex. 44.

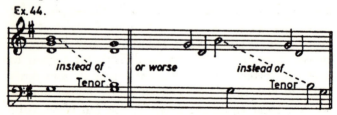

If, in addition to being an accompanist for a choir,

you are also its deputy conductor, you may from time to time be asked to take charge of a rehearsal. Once again I suggest very strongly that you do nothing which may undermine the authority of the conductor on his return. Confine your rehearsal to consolidating the work of previous rehearsals; or, if you are asked to rehearse new music, make sure that you know how the conductor wants it to be performed. You can do no better than read and act upon the suggestions given by Harvey Grace in *The Complete Organist* (Chapter 18).

It may be that there is no conductor, and you are conductor as well as accompanist. You must place yourself in order to be seen by as many of the choir as possible. Again it is much easier with a grand piano, which you can place either at the side or (my own preference) in the middle of the platform, so that you are in the conductor's usual position with your back to the audience. The best position with an upright piano is at the side of the platform, with your right hand nearer to the choir.

When you are accompanist and conductor your choir lacks the confident direction that a conductor

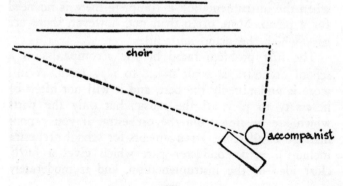

with both hands free can give them. You can help them to some extent by conducting with your right hand and playing with your left, but this is a task for experts and it needs keen co-ordination. An expressive face can assist at times with difficult leads, sudden *pianissimi,* and the like. These aids, however, will be available only spasmodically, because playing will take most of your attention. It is in fact by your playing that you can give back to your choir some of the confidence it has lost by being without a conductor. In order to do this, you will often have to play with more marked rhythm and perhaps slightly less smoothly than you otherwise would, and you may have to shorten chords at the ends of phrases in order to help them to breathe. Do not, however, exaggerate these differences. A choir needs above all practice in singing without a conductor, and help from an accompanist must be given with good taste and finesse.

### PLAYING IN AN ORCHESTRA

*School orchestras*

More and more schools have a full orchestra, and when the instrumentation is complete there is no need for a piano. More often than not, however, there are gaps which the piano must fill.

The first problem faced by the accompanist in a school orchestra is what music to play from. A full score is undoubtedly the best, and it will not often be necessary to play all the music, but only the parts which are missing from the orchestra. If you cannot manage a full score, arrangements for school orchestra include a piano-conductor part which gives a fairly clear idea of the instrumentation, and is moderately easy to read. This is a useful substitute, from which you must then decide what to play and what to leave out.

To play the piano in a school orchestra calls for a similar attitude to the *continuo* player's. It will more often than not be a case of filling in the middle between a large number of violins and a heavy bass. Yet nearly always this is what one hears: an already overloaded melody being reinforced on the piano, the main effect being to point out clearly how much out of tune the violins are playing.

(This is a matter which also concerns the song-accompanist. Schumann, to mention no other composers, is fond of doubling the voice part on the piano. It is unnecessary to thump this out when the singer is singing in tune, and it is most distressing to do so when he isn't. It merely shows the audience, most of whom might otherwise not know, how flat he is singing.)

Most school orchestras have a cello or two, and possibly a trombone. Therefore to play the bass line at that pitch is superfluous. Not so many schools, however, have double-basses, in which case it will be necessary to play the bass at that pitch—an octave lower than the cellos. Unless there is plenty of horn and viola support, the middle will need to be reinforced on the piano.

If one instrument has a solo, resist the temptation to double it on the piano, however inexperienced the player may be. It is much better for him to make his mistakes unaided, and by learning to play without assistance he will gain confidence. Only when disaster is imminent is it excusable to come to the rescue.

There may be an oboe player who can be relied

upon to play an unwavering note, in which case let him play the tuning A, having first made sure that he is in tune with the piano. This helps in two ways: it gives him practice in a duty he will later have to perform, and it gives the others practice in tuning to the sound they will hear when they play in an adult orchestra. String-players especially need this practice, in order that they can learn to tune their own fifths and not take them from the piano.

## Professional orchestras

Playing an orchestral keyboard part, whether piano, organ or celesta, is quite a different matter, and obviously you must play what the composer has written and nothing else. In any case, the conditions are likely to be very different, because you will almost certainly be playing with professional instrumentalists. Not many amateur bodies can tackle Stravinsky, Debussy, and other composers who occasionally write a piano part in their orchestral works. Here I offer one warning. Professional orchestral players often play behind the beat, especially in slow music. I have never discovered why they do this, and indeed if one asks them individually about it one receives a number of different answers. Nevertheless, play behind the beat they do, and this can be most disconcerting to one unused to it. It may be that you will become accustomed to it quickly, but if not I suggest that you stop looking at the beat and listen to the music around you instead. In any case, listening as carefully as possible to the players near you helps a great deal in playing your part. In this way, too, you can soon give up the chore, so tiresome to a pianist, of counting bars' rest.

# 13

# THE ACCOMPANIST IN SCHOOL

Much of the work of the accompanist in school will consist in playing for choirs, and this matter was dealt with in the last chapter. At the same time, there are many other duties which will concern him.

I shall deal first with playing for public gatherings, morning prayers, speech days, and so on. In playing for a large body of singers there can be little thought of accompanying them. It is much more a matter of conducting by playing. Your playing will need to be positive and forceful; it will frequently be in front of the singing (especially if there has been no rehearsal beforehand). Its rhythm must be relentless, if the pace is not to become slower and slower. (I have no doubt that many church organists have before them at the beginning of their hymns the direction *Andante, poco a poco rall al fine,* which may be six verses away).

There is no need for these gloomy suggestions at morning assembly or prayers, because presumably there will be a regular rehearsal of the hymns which are to be sung then. To go minutely into the technique of hymn-practising is beyond the scope of this book, but there is one matter of concern to the accompanist, and that is the beginning and end of verses. If these can be made tidy it makes a vast difference to the effectiveness of the singing. There are a number of different ways in which this tidiness can be achieved.

One of the easiest is to arrange to play the opening chord one beat before the singing is to begin, and to hold it (that is, not play it again) during the first note of the singing. This gathering chord is frowned upon by many, but I see no harm in it. It is certainly better than playing the melody-note or bass-note one beat in advance. It is easily rehearsed, and understood by all. Moreover, it has the advantage that it is possible to alter at will the length of the pause between verses.

Another method is to arrange for a definite number of beats between the verses, and between the playing-over and the first verse. This method commends itself to many choirmasters, because it is regular. Any method which is effective will do, but these two are probably the best.

As far as the end of the verse is concerned, there is no difficulty with organ accompaniment, because the singers can stop as soon as the organ stops, and this (with rehearsal) produces a tidy ending. With piano it is rather more difficult, and once more the best method is probably to count beats, for example

Ex.45.

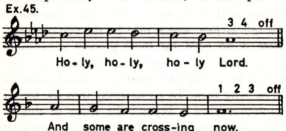

This will probably be the normal method even with organ, but it may be that a longer pause is required for the end of the last verse.

With the question of Amens to hymns I shall not

deal. I only ask, if Amen is to be sung, that it should grow naturally from the last verse and should not be sung and played *mezzo piano* after a *fortissimo* last verse, that the singers should know how many beats each of the two notes is to last, and that the following sort of performance should be buried for ever.

Ex. 46.

*God save the Queen*

No accompanist can so call himself unless he can play the National Anthem by heart. Here I make only two suggestions.

(a) Do not regard G as the only possible key. A girls' school, for example, would sing it much more effectively in A or even B flat. A meeting of parents might more conveniently use F, and a governors' meeting (I speak as a governor) possibly D flat.

(b) Play the warning chord firmly, but quietly enough for nobody to mistake it for the beginning. Leave time for all to take breath, and then begin *fortissimo*. On no account wait on the first chord.

So much for the multitude. Now for the individual.

Playing for individual children can be a very great

pleasure. It is possible to help them enormously, and in the case of instrumentalists to supplement the technical training given by their professional teachers.

*Singers*

These may not be having private lessons with a professional teacher, in which case the schoolmaster has a heavy responsibility. He must first of all be sure that no music that he gives the pupil to learn will impose any strain on the voice. The boy's breaking voice presents problems of its own with which I am not at the moment concerned. Broadly speaking, however, any voice aged between fourteen and eighteen needs care. Do not be beguiled by the thoughts of great applause on Speech Day into asking a girl of fifteen to sing 'One fine day'.* Unless the voice is an exceptional one, the strain of rehearsal will lay up great trouble for it in the future. Do not force any pupils to sing beyond their natural capabilities in tone and range. Be content with music which does not stretch them to the absolute limit.

At the same time, acquaint young singers as much as possible with the best music. There are plenty of arias by Handel, and songs by Schubert, Brahms, Warlock, and Vaughan Williams (to mention no others), which young boys and girls can sing. Of course they will not give brilliant accounts of such music, but to let them have the chance to learn it is far preferable to fobbing them off with third-rate festival pieces.

Try to make them sing intelligently. Good articula-

* I know Butterfly was only fifteen, but you don't ask only consumptive girls to sing Mimi.

tion is essential, but it is not enough. A singer, young or old, must feel the meaning behind the words if he is to interpret successfully.

Good phrasing depends upon good breathing. Do not let them hunch their shoulders while breathing. Make them feel that the important part of breathing is located immediately under the ribs.

Your accompaniment should guide them tactfully. It should not be domineering, but should give assistance in subtle ways, slightly moving the music forward when immature breath-control fails, perhaps giving them fractionally longer to take breaths between phrases (to be able to snatch quick breaths successfully is one of the marks of a mature singer), and in general giving good, sustained support. Do not make the mistake of under-playing the accompaniment for an undeveloped voice; it probably needs more support than an adult's.

## Instrumentalists

It is better on the whole not to interfere with the technical problems of young instrumentalists, and if you have confidence in their teachers you will not want to do so. When playing for them, confine your help to making better musicians of them.

Nag them about intonation. Teach them to recognise the difference between sharp and flat playing (you may first of all have to teach yourself). This is not so hard as it may sound. Some can do it naturally better than others, but all can improve with practice.

It is impossible to describe what sharp and flat playing and singing feel like to the listener. My own reactions are that when I hear something flat I want to tense my muscles (especially the ones in my head), hoping, I suppose, that this will help to raise the pitch. When I hear something sharp I open my throat as wide as I can (this is almost instinctive, and I have usually done so before I am aware of it). These are not so much feelings as reactions, and I need hardly say that neither of them ever has the slightest effect upon the pitch.

Beginner string-players tend in my experience to play flat when using the third and fourth fingers. Show them how a small movement of the finger makes a large difference to the pitch. My impression is that young string players correct by far too much.

Wind players have quite different intonation problems.* There are certain notes on woodwind instruments which are naturally out of tune, and need care in blowing (the pitch of a woodwind note can be altered by blowing it differently or by changing the lip-pressure).

(1) On the flute the two lowest notes 46a are inclined to be flat, and those an octave higher 46b sharp. In the top octave the tendency is again (though somewhat less) to be flat on 46c.

Ex. 46

(a)                    (b)                    (c)

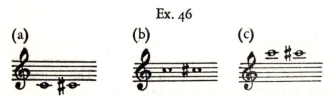

(2) About the oboe it is a little more difficult to be precise, because one instrument differs very much from

* I am very grateful to some of my orchestral colleagues in Belfast for assistance with this section.

another (this is to some extent true of all woodwind instruments). However, it can be said that the same tendency towards flatness can be discerned on an oboe's lowest notes, especially 46d, and that there is a general inclination towards sharpness on the high notes from about 46e upwards.

(3) The clarinet has a patch between 46f (sounding 46g) on the B flat clarinet in which the notes are of

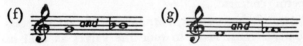

less good quality and inclined towards sharpness. It has been suggested that concert A is not the best note for a beginner clarinet-player to tune to, because it uses the whole length of the instrument, and therefore is liable to be even more flat than the others when the instrument is cold. A better note may be 46h, sounding 46i

on the B flat clarinet. Of course, when the player joins an orchestra he will have to tune to A like everyone else.

(4) The notes 46j on the bassoon, and to a lesser extent 46k, are often sharp.

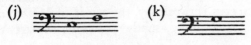

In general, the better the quality of the instrument, the more in tune will it play. Woodwind instruments are much affected (sharpwards) by heat, and all players must re-tune when their instruments have warmed up. The second half of an orchestral concert seldom takes place at the same pitch as the first half, and the matter is aggravated when a piano is involved. What usually happens is that the piano A is played and the woodwind underblows its A in order to be in tune. The music then starts; the woodwind blows normally and is sharp. I never remember not wincing at the entry of the piano in the first movement of the 'Emperor' concerto.

Players of the trumpet, horn, and other valved brass instruments must learn to tune the valves as well as the main tuning-slide. To take an example, the note D on a B flat trumpet, written E, should sound at

(1)

the same pitch, whether it is played open, or with the first and second valves, or with the third valve. Brass instruments can also vary the pitch of a note by alteration of the lip and breath-pressure.

Apart from this important question of intonation, young players need much help in matters of style and phrasing, which are closely related. Style is a quality scarcely definable, and one very difficult to communicate to young people. It consists first of all in recognising that hardly any two composers should be performed in exactly the same way, that there are great differences between say, the music of Mozart and Beethoven, and more subtly between Mozart and

Haydn. Style can be more easily pointed out in the particular than in the general, for applied style takes notice of such details as comparative climaxes (not all *fortissimi* are of the same loudness, even in one piece or movement), the comparative length of notes (some crotchets are longer than others, even in one piece or movement), and phrasing.

Phrasing can be discussed more concretely, but even here details are more easily described than generalities. For instance, the maxim that all phrases are curved, with the climax at the top of the curve, needs a great deal of qualification before it can be accepted. If the implication is that the curve is of the conventional rainbow shape, the maxim is quite unacceptable, because it has too many exceptions. Some phrases have their climax at the end, some at the beginning, and some have no climax at all.

Instruction in the playing of the ends of phrases needs to be given with great care. The clipped phrase-ending is usually thought to be inartistic; and often it is, but not always. If one phrase is immediately followed by another, it is most inartistic *not* to shorten the last note of the first phrase; especially in quick music, because either there is no break and the phrases are run into one, or the second phrase begins late. Management of this type of phrase-ending is easier in one sense for singers and wind-instrumentalists because they have to breathe. Yet for the same reason it is more difficult, because they often have to end a phrase, breathe, and begin a new one, in a very short space of time. String-players, however, have to be told to make a break, either by lifting, or by stopping the bow on the string. The important thing to show young people is that, if the phrase ends off the beat and the last note is short, it must be unstressed and not given a *staccato* accent. In Ex. 47 (from Schubert, *Rosamunde* overture) the quavers should be unstressed.

Ex. 47.

Even if it is impossible to go deeply into these questions with young players, it is certainly worth while to bring them to their notice. In the early stages it may be sufficient to show a pupil that notes are to be joined together to make a phrase, and not to be played as separate entities. Later, the more subtle questions of which note in a phrase is the loudest and whether the speed changes at all during a phrase can be introduced.

In giving all this help there is the danger of stifling the adolescent personality. Few events are more gloomy than a Speech Day concert in which all the solo items are heavily stamped with the character of the director of music, with never a chance for the young performer to project his own character, however immature. Let the young soloist make his own interpretative mistakes, but do let him perform, and not merely mimic. This needs a really fine balance of control from the schoolmaster, trying to instil good taste, and at the same time helping the teenager to develop a musical personality of his own.

Finally, the school music-teacher has a responsibility

to give encouragement and help to budding accompanists. It is not difficult to pick them out; they will almost certainly be interested in more than one kind of music, as opposed to the soloists, who may have narrower interests; and they will probably be keen members of the school choir and amongst the best sight-singers in it. Above all, when given a chance to play for someone else, they will show sensitivity and flexibility even if their technique is not first-class.

They should be given every opportunity to arrange chamber music, if there happen to be some promising string-players, and to coach solo singers for a concert; and eventually to play for a local singing-teacher. I am sure that almost all the best accompanists would say that they began in this way, and that they drifted almost imperceptibly into professional accompanying.

## APPENDIX A
## SUGGESTIONS FOR FURTHER READING

**Books**

FERGUSON, H.    Style and Interpretation, an Anthology of keyboard music (especially the Introductions to volumes 1 and 2): Oxford University Press, 1963.

GRACE, H.    The Complete Organist; the Richards Press Ltd., 1920.

JACOB, G.    Orchestral technique; Oxford University Press, 1931.

MOORE, G.    The Unashamed Accompanist; Ascherberg, Hopwood and Crew, 1943.
Singer and Accompanist; Methuen, 1953.
Am I too Loud? Hamilton, 1962.

MORRIS, R. O.    Figured Harmony at the keyboard; Oxford University Press, 1932.

MORRIS, R. O. and FERGUSON, H.    Preparatory Exercises in Score-reading; Oxford University Press.

*Articles and Lectures*

BERGMANN, W.    Some old and new problems of playing the *basso continuo;* Proceedings of the Royal Musical Association, 87th Session, 1960/61.

*Grove's Dictionary of Music and Musicians;* 5th edition, Macmillan, 1954. Articles on 'Harpsichord playing' by R. Donington, and on 'Thorough bass' by F. T. Arnold and R. Donington.

*Hinrichsen's Musical Year Book, Volumes IV-V,* 1947-48. Article, Recent researches in score-reading, by H. Lowery.

## APPENDIX B

## A MINIMUM REPERTORY FOR AN ACCOMPANIST

The accompanist's repertory is never large enough. There is always new music to be learnt, which is partly what makes accompanying such an interesting career or pastime. The list below shows what I consider to be a minimum for anyone who has begun to play regularly as an accompanist. The player of some years' experience will know most of it already, together with much that is not included in it. But if it reveals that there are gaps in your own repertory, begin to fill them. The list should be regarded as a point from which to branch out in all directions.

I have specified composers rather than works, because I feel that, for instance, if you know a violin sonata by Mozart, you are at least part of the way towards knowing another. A work may be considered to be in your repertory if you could play a performance of it at a day's notice for a song, and a week's notice for a chamber work, practising part-time.

(1) Two hundred songs; consisting of one hundred by Schubert, Schumann, Brahms, and Wolf; and one hundred by other composers such as Fauré (and other French composers), Parry, Stanford, Gurney, Vaughan Williams, Warlock, Britten, etc. The second hundred may include the commoner folk-song arrangements.

(2) At least one violin sonata by each of the following composers: Handel, Mozart, Beethoven, and Brahms.

103

(3) Any two of the piano quintets of the following composers: Schubert, Schumann, Brahms and Dvořák. You may not have many opportunities to perform these.

(4) Fifty operatic and oratorio arias by Bach, Handel, Haydn, Mozart, Verdi, Puccini, etc.

(5) If you are accompanist to a choral or operatic society, your repertory will be dictated by what the society performs. You should, however, be able to rehearse any of the following works at fairly short notice: *Messiah, St. Matthew Passion,* the *Requiems* of Mozart, Brahms, and Fauré, together with a number of operettas by Johann Strauss, Lehár, Sullivan, etc.

# GENERAL INDEX

Accompaniment, choral, 84–88
  congregational, 68–70
  discreet, 7
  harpsichord, 61–62
  in school, 88–90, 91–101
  orchestral, 88–90
  organ, 63–70, 92
  young singers, 94–95
Accompanist, assertive, 9
Alto clef, 53
Amen, 92–93

Balance, 28–29, 63, 74–79
Blend, 79–80

Cadences, 41–42
Chamber music, 71–83
Choirs, playing for, 84–88
Clef, alto, 53
  tenor, 54
Conductor, accompanist's attitude towards, **85**
Continuo, 57–62, 72–73
Continuo reduction, 48–50

Encores, 28, 34

Harpsichord accompaniment, 61–62

Instrumentalists in school, 95–101

Nerves, 31, 33

Orchestra, playing in a professional, 90
  playing in a school, 88–90
Orchestral scores, 53–56
Organ accompaniment, 63–70, 92
  reductions, 65–68

Pedal, left-hand, 23–24
  sustaining, 19–23
Pedalling, 19–24
Pedals, organ, 64, 65

105

# INDEX OF WORKS MENTIONED
## IN THE TEXT

107